SURVEYING MATHEMATICS MADE SIMPLE

An original Book by

Jim Crume P.L.S., M.S., CFedS

Co-Authors
Cindy Crume
Bridget Crume
Rebecca Crume
Troy Ray R.L.S.
Mark Sandwick P.L.S.
Mark Lull

KINDLE - PRINTED EDITIONS

*****

PUBLISHED BY:

Jim Crume P.L.S., M.S., CFedS

Geodetic Datums Made Simple

First publication: September, 2016

Cover photo: Vertical Control Point
Camp Verde, Arizona

TERMS AND CONDITIONS

TABLE OF CONTENTS

INTRODUCTION

Straight forward Step-by-Step instructions.

This book is just one part in a series of digital and paperback books on Surveying Mathematics Made Simple. The subject matter in this book will utilize the methods and formulas that are covered in the books that precede it. If you have not read the preceding books, you are encouraged to review a copy before proceeding forward with this book.

For a list of books in this series, please visit:

http://www.cc4w.net/ebooks.html

Prerequisites for this book:

A basic knowledge of geometry, algebra, and trigonometry, are required for the explanations shown in this series of books.

Operating survey equipment and field procedures will not be covered in this book.

The intricate details of geodesy such as latitude, longitude, ellipsoid models, etc. are not covered in this book. There is plenty of information available should you want to dive deeper into that subject. This book is simple in that it covers the bare minimum needed for construction, land boundary and highway surveys.

The following books of this series are recommended to assist with the processes mentioned in this book:

4

Bearings and Azimuths - Book 1
Create Rectangular Coordinates - Book 2
Inverse Between Rectangular Coordinates - Book 3
Circular Curves - Book 4
Parcel Boundary - Book 5
Coordinate Transformation - Book 9

These books contain formulas, step by step solutions and examples.

 Throughout this book, tips will be given to help explain or give directions on the subject matter.

DEFINITIONS

ASCII: American Standard Code for Information Interchange. The file is a character-encoding scheme. Originally based on the English alphabet, it encodes 128 specific characters into 7-bit binary integers. For the purposes of this book, an ASCII file refers to a text file that contains the project Cartesian/Rectangular coordinates in a "P,N,E,Elev,Desc" format. Simple text programs such as Windows Notepad or Mac TextEdit can be used to open/edit this file type. [Point#, Northing, Easting, Elevation, Description] (i.e. 1000, 1200.00, 1400.00, 100.00, Point A)

Coordinates: Coordinates are points that represent a position of a point whose values are fixed either by a found monument or calculated position to a defined datum.

Geodetic System: Is a coordinate system, and set of reference points, used to locate places on the Earth.

Geoid: The hypothetical shape of the earth, coinciding with mean sea level and its imagined extension under (or over) land areas.

Grid Adjustment Factor (GAF): Also known as the Combined Factor. This factor is a combination of the Grid Factor and Ellipsoid Factor for converting from Grid Coordinates to Ground Coordinates and vice versa.

Datum-Horizontal: The horizontal datum is used to measure positions on the earth surface from a standard reference point.

Datum-Vertical: The vertical datum is used as a reference point for elevations of surfaces and features on the earth.

Ellipsoid: A mathematical model of a closed quadric surface that is a three-dimensional analogue of an ellipse.

Latitude: The angular distance of a place north or south of the earth's equator usually expressed in degrees, minutes and seconds.

Longitude: The angular distance of a place east or west of Greenwich Meridian usually expressed in degrees, minutes and seconds.

Mean Sea Level: An average level for the surface of Earth's oceans from which elevations are measured.

NAD 27: The North American Datum of 1927 (NAD 27) is the horizontal control datum for the United States that was defined by a location and azimuth on the Clarke spheroid of 1866, with the origin at Meades Ranch, Kansas.

NAD 83: The North American Datum of 1983 (NAD 83) is the horizontal control datum for the United States, Canada, Mexico and Central America, based on a geocentric origin and the Geodetic Reference System of 1980.

NGVD 29: The Sea Level Datum of 1929 was named the National Geodetic Vertical Datum of 1929 on May 10, 1973. The datum was used to measure elevation above or below mean sea level.

NAVD 88: The North American Vertical Datum of 1988 (NAVD 88) is the vertical control datum established in 1991 by the minimum-constraint adjustment of the Canadian-Mexican-United States leveling observations.

NGS: The National Geodetic Survey (NGS), formerly the *United States Survey of the Coast* (1807 - 1836), *United States Coast Survey* (1836 - 1878), and *United States Coast and Geodetic Survey* (1878 - 1970), is a United States federal agency that defines and manages a national coordinate system.

OPUS: The Online Positioning User Service provides simplified access to high-accuracy National Spatial Reference System (NSRS) coordinates.

Projection-Lambert: A Lambert conformal conic projection is a conic map projection used for State Plane Coordinate System, and many national and regional systems.

Projection-Low Distortion: A custom mapping projection, which can utilize Lambert, Transverse Mercator or other projection models, to minimize the difference between ground and grid distances.

Projection-Transverse Mercator: The Transverse Mercator projection is a cylindrical map projection used for State Plane Coordinate System, and many national and regional systems.

WGS 84: The World Geodetic System (1984) is a standard for use in cartography, geodesy, and navigation including GPS.

BACKGROUND INFORMATION

My surveying experience began in 1973 while attending Central New Mexico Community College (formally known as Albuquerque Technical Vocational Institute). I have since then became registered in several states, appointed as a Mineral Surveyor and became a Certified Federal Surveyor.

My experience includes most all aspects of surveying except underground and hydrographic.

Since 1990, my primary survey experience has been highway right-of-way retracement as an on-call for the Arizona Department of Transportation (ADOT). During this time frame I have retraced hundreds of miles of highway right of way. I have prepared numerous Right of Way Plans and Results of Surveys to ADOT standards. I also worked with the ADOT CADD committee to develop the ADOT Right of Way CADD Standards.

I have had the opportunity to work closely with the ADOT Right of Way Project Development Manager over the years in analyzing centerline alignments to "Best Fit" the controlling Right of Way centerline and corridor with existing monumentation.

I also have experience in determining right of way corridor's with Maricopa County Department of Transportation, Maricopa County Flood Control District, Mohave County Flood Control District, Gila River Indian Community, Salt River Pima-Maricopa Indian Community, Navajo DOT and other public agencies.

One of the primary components for all the surveys that I have performed over the many years is the controlling datum. All surveys require a horizontal datum while some will also require a vertical datum depending on the survey project.

Ok, enough on the disclaimer and notices. Let's get on to the fun stuff.

RELATIVITY

For any survey project, the geometric relationship between all points within the survey will be a constant. Their relativity to each other does not change regardless of the datum. This is true for both horizontal and vertical relationships.

I am not talking about the "General Theory of Relativity" by Albert Einstein, but the geometric relativity between points of a survey.

The relationship is based upon the how points to be surveyed are shown on a plat, legal description, construction plans, etc.

The Angular and Distance relationship between points will always be the same as well as the vertical differences regardless of the datum.

This geometric relationship between points is constant regardless of the number of points.

Below is a simple parcel boundary that shows the angular and distance relationship between the points.

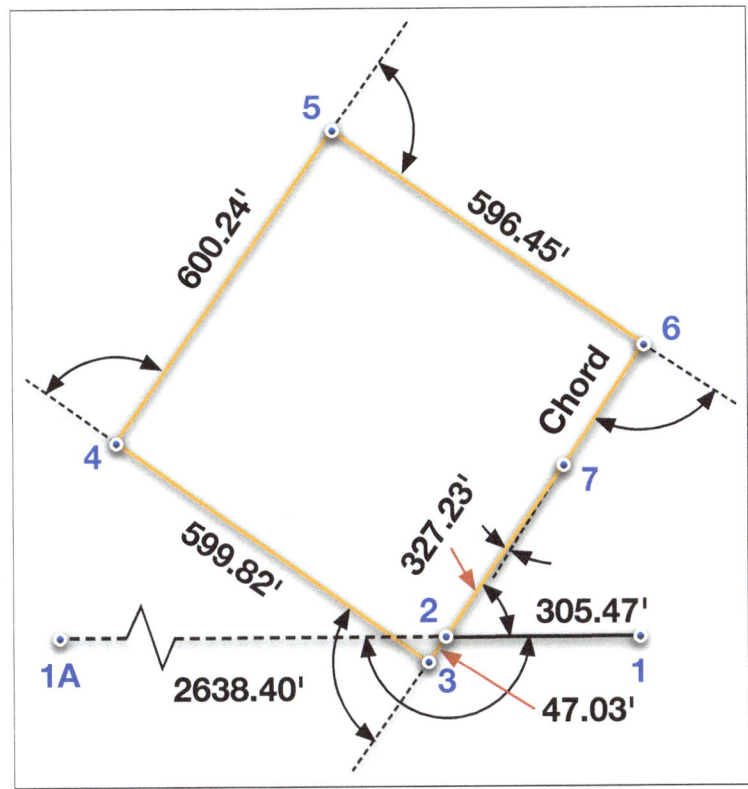

The points in the parcel can be moved, rotated, raised or lowered as a group to any datum and the geometric relationship between the points will not change.

Once you grasp that concept, then the datum becomes secondary to the survey. The primary factors are the geometric relationship between the points via angles, distances and elevation differences.

WHAT ARE GEODETIC DATUMS?

In surveying, a datum is a standard reference point from which survey measurements are made.

11

For every project site, there will be a Base Point (Standard Reference Point) or set of Base Points depending on the size of the project. The Base Point datum can be assumed (GPS Autonomous), assumed (Local coordinates), Public Agency defined, NGS (State Plane Coordinates), modified State Plane Coordinates or custom (Low Distortion Projection, LDP).

Which datum you use will depend upon the project and the client requirements.

For example: On ADOT projects, the field work is performed using NGS State Plane Coordinates (Grid). The deliverables are Grid Bearings, Ground coordinates and Ground distances on the Results of Survey and Final R/W Plans.

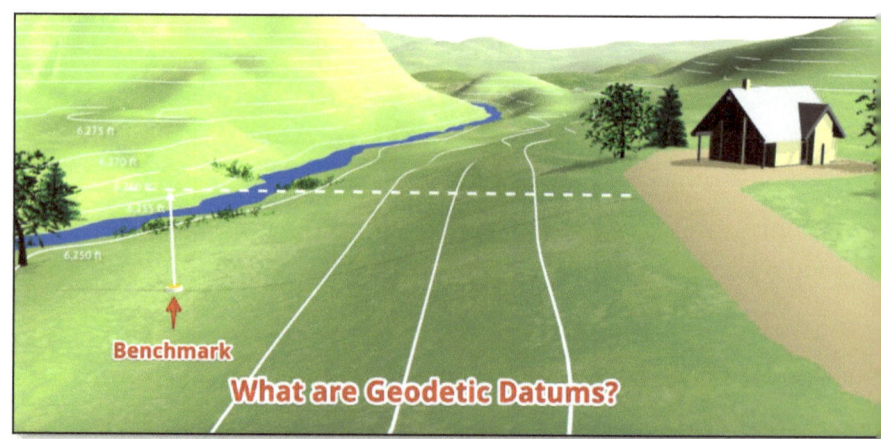

This YouTube Video "What are Geodetic Datums" is an excellent illustration of what datums are.

https://youtu.be/kXTHaMY3cVk

BEARING DATUM

Modern surveys are performed with GPS Real Time Kinematics (RTK), therefore Grid Bearings are required to be used as the Bearing Datum.

Preliminary search coordinates are prepared in ground coordinates using legal descriptions, plats, results of surveys, etc. which are almost always in ground distances.

Bearings on record documentation will need to be rotated to Grid Bearings before being utilized to generate preliminary search coordinates for the field crews.

The bearing datum on record documents are usually unknown especially for older documents. They can be Grid, Geodetic (True), Magnetic or assumed.

When I begin creating search coordinates, I identify a line that is common among the record documentation (usually a section line) and determine the Grid bearing for this line to rotate the record documents to so that the coordinates are on the same bearing datum. Further adjustments from ground distances to grid distances will also be required.

 I go into great detail on how to determine the Grid bearing for a project as well as creating search coordinates that are very close to Grid coordinates in my book "Highway Centerlines (Retracement) - Book 14".

See below for a typical Basis of Bearing note:

"The Basis of Bearings for this survey are Grid per NAD 83 as adopted by the Gila River Indian Community Council per Resolution GR-18-97 Survey Control Network."

The convergence angle from Grid North to Geodetic North i Zero at the Central Meridian. The convergence angle increases as you move further left and right of the Central Meridian. The convergence angle is negative left of the central meridian and positive right of the central meridian a shown on the NGS Data Sheets.

Conversion from Geodetic North to Grid North

Examples:
Convergence (-0 07 43.6) left of central meridian

NE Quadrant
N 45-00-00 E - (- 0 07 43.6) = N 45-07-43.6 E Grid

SE Quadrant
S 45-00-00 E + (-0 07 43.6) = S 44-52-16.4 E Grid

SW Quadrant
S 45-00-00 W - (-0 07 43.6) = S 45-07-43.6 W Grid

NW Quadrant
N 45-00-00 W + (-0 07 43.6) = N 44-52-16.4 W Grid

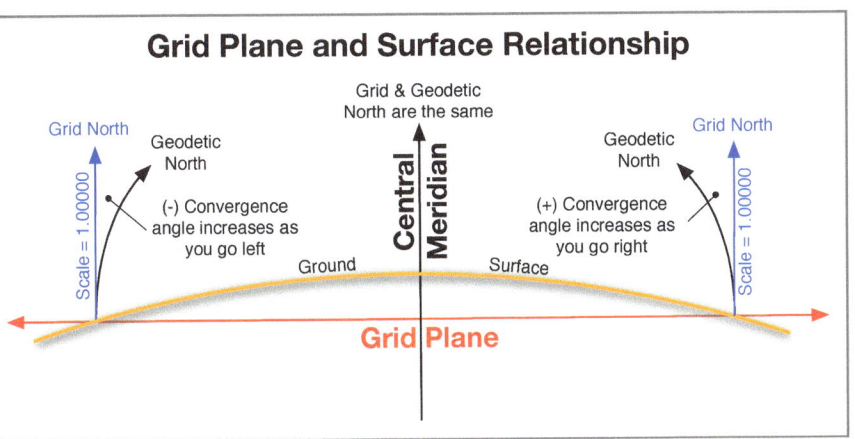

BASE POINT

A Northing coordinate, Easting coordinate and Elevation (Meta Data) will need to be established for each Base Point for the survey project. For some projects, a Latitude, Longitude and Ellipsoid Height will also be required especially if you are working on a Grid datum.

These Meta Data values are the datum that will be used for the duration of the project. All survey measurements will be performed from the Base Point.

The field procedures to establish the Meta Data for the Base Point will vary depending on the survey equipment that is used (GPS, Total Station, Direct Levels, etc.) and the client requirements.

STATE PLANE COORDINATES

The remainder of this book will explain State Plane Coordinates System (SPCS) in simple terms for use on land and highway survey projects. The SPCS is directly connected to the NGS National Spatial Reference System.

The SPCS is a set of geographic zones designed for specific regions of the United States. Each State contains one or more state plane zones. Zones are either a Transverse Mercator Projection or Lambert Conformal Conic Projection. The choice between the two map projects is based upon the shape of the state and its zones.

Generally a Zone that is longer north and south will usually be a Transverse Mercator Projection system. A Zone that is longer east and west will usually be a Lambert Mercator Conic Projection system.

Below are the SPCS Zones for the USA.

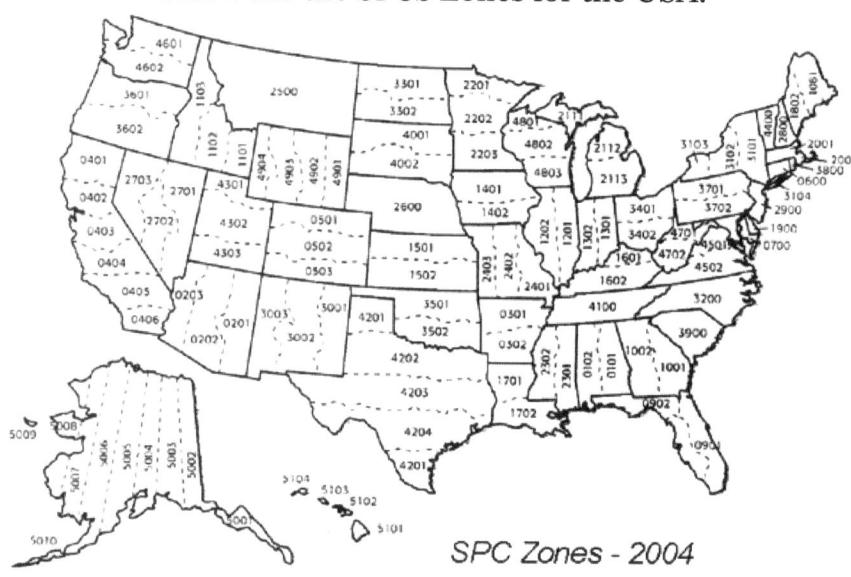

SPC Zones - 2004

Other countries also have established projection systems that are specific for their region.

There is also a Universal Transverse Mercator system with multiple zones that covers the entire world as shown as follows:

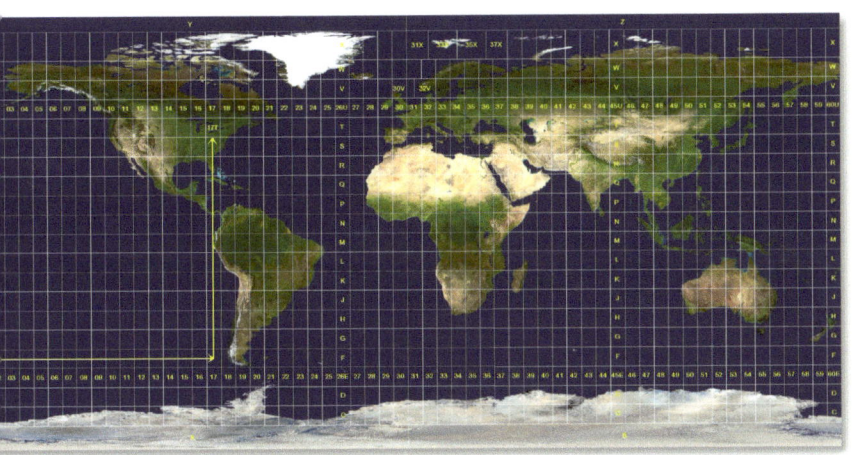

A Low Distortion Projection is a custom system that is designed for a specific location that minimizes the distortion (difference) between Grid and Ground distances. This projection is a local system and is not part of the SPCS.

The mathematics for each of these projections is quite complex. Fortunately there are software solutions that take care of the complexity of the calculations.

Regardless of which map projection system is used, it needs to be well documented on a survey so that it can be retraced by others.

Below is an example of the documentation for a typical Results of Survey.

GENERAL NOTES

THREE P&M JOB NO. 3830 CONTROL STATIONS WERE USED TO ESTABLISH THE
CONTROL BASELINE FOR THIS SURVEY AND THEY ARE AS FOLLOWS:

Z 284 NAD 83(2007) 33°03'39.14755"(N) 112°03'19.10463"(W)
Grid Coordinates N 749706.94989 E 657525.66990
Ground Coordinates (Pt. 104) N 749826.90300 E 657630.87400

KIM-LEL NAD 83(2007) 33°03'09.78314"(N) 112°01'54.14040"(W)
Grid Coordinates N 746730.56011 E 664752.96953
Ground Coordinates (Pt. 111) N 746850.03699 E 664859.33000

HC 171.5 NAD 83(2007) 33°01'43.63030"(N) 112°02'51.33429"(W)
Grid Coordinates N 738029.41030 E 659874.41010
Ground Coordinates (Pt. 112) N 738147.49500 E 659979.99000

THE FOLLOWING PARAMETERS WERE SET FOR THE BASIS OF THIS SURVEY:

SYSTEM: UNITED STATES STATE PLANE 1983
ZONE: ARIZONA CENTRAL 0202
DATUM: NORTH AMERICAN 1983 (CONUS) EQUIVALENT TO WGS-84
GEOID MODEL: GEOID03 (CONTINENTAL US)

UNITS = INTERNATIONAL FEET (1 FOOT = 0.3048 METERS EXACTLY)

DIRECT SATELLITE MEASUREMENTS WERE MADE ON EACH POINT ON THE
NORTH AMERICAN DATUM OF 1983 WITH AUTOMATIC TRANSLATIONS MADE
TO THE STATE PLANE GRID SYSTEM.

A GRID ADJUSTMENT FACTOR OF 1.00016 AND SCALE ORIGIN VALUES OF
N=0.00, E=0.00, AS OBTAINED FROM ADOT (P&M JOB NO. 3830) WERE USED FOR THE
TRANSLATION TO THE SURFACE. GRID COORDINATE "X" GAF = SURFACE/
GROUND COORDINATE.

ALL BEARINGS ARE GRID BEARINGS, DISTANCES ARE GROUND DISTANCES AND
COORDINATES ARE GROUND COORDINATES.

THE EXISTING ADOT R/W SHOWN ON THIS RESULTS OF SURVEY WAS BASED
ON AN OBSERVATION AND EVALUATION OF PHYSICAL CORROBORATING EVIDENCE
IN THE FIELD AND A REVIEW AND ANALYSIS OF EXISTING R/W PLANS, MAPS AND
R/W TITLE DOCUMENTS AVAILABLE AT THE TIME THE SURVEY WAS PREPARED;
AND MAY NOT REFLECT PRIOR OR PENDING ADOT DISPOSALS OR ACQUISITIONS
THAT WERE NOT SHOWN ON SAID PLANS AND DOCUMENTS AT THE TIME THE
SURVEY WAS PREPARED.

THIS RESULTS OF SURVEY IS ON FILE IN THE FILE ROOM OF THE RIGHT OF WAY
PLANS SECTION OF THE ARIZONA DEPARTMENT OF TRANSPORTATION, 205 SOUTH
17TH AVE, MD 612E, PHOENIX, AZ 85007. (602) 712-7270 SEE ADOT R/W FILE
COPY FOR MOST CURRENT ADDITIONS OR REVISIONS.

FOR ADDITIONAL INFORMATION ON SECTION CORNER DESCRIPTIONS, SEE
CORNER RECOVERY SHEETS IN THE ADOT R/W PLANS SECTION RECORDS.

There are four options available to establish Meta Data for
the project Base Point.

1. Locate the Base Point from a known control point. For some projects, the known control point will be used as the Base Point if it is located in the project area.

2. Collect GPS Static data on the Base Point and process through baseline software or through OPUS Projects.

3. Use a GPS Real-Time Network to create a Virtual Base Point.

4. Set a total station in a random location and perform a resection to two or more known points.

In most cases, GPS will be used to establish the Base Point control. For high precision vertical control, direct levels will be run from a known benchmark.

On construction sites and projects that require high precision, a Total Station will be utilized on the Base Point and direct levels will be ran as needed to check the precision of the Total Station. A minimum of two Control Points are required for a Total Station setup. One to set the Total Station on and the other for a back sight.

For large scale projects such as land boundary and highway surveys, a GPS RTK system should be utilized.

Once the Base Point has been established, measurements (vectors) will be made from the Base Point to each point in the survey.

GPS VECTORS

As of the writing of this book, World Geodetic System (WGS) 84 is used by GPS satellite receivers to measure the Latitude, Longitude, Ellipsoid Height and Time at the Base Point and simultaneous readings at the Rover.

A UHF radio or cell phone connection is used to send the Base Point information to the rover and the data collector.

The software in the data collector uses the WGS 84 position of the Base Point and Rover and converts it to the map projection model that is assigned to the project (i.e. Arizona Central Zone). The Grid bearing, Grid distance and Vertical difference between the Base Point and the Rover position are calculated using RTK mathematical solutions. Grid coordinates and elevation values are calculated for the point being measured using the GPS vectors.

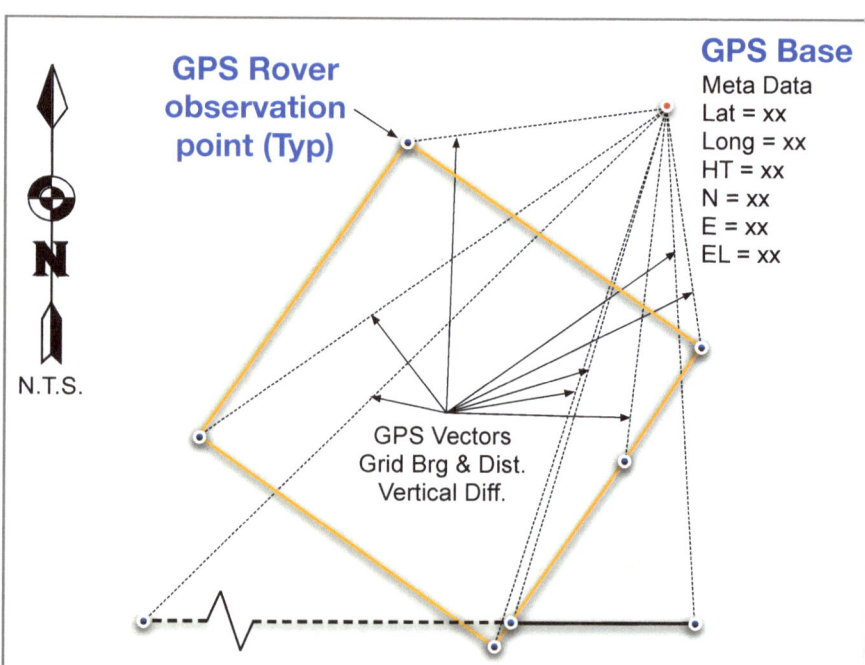

The Grid Northing and Easting coordinates will need to be converted to Ground coordinates with a project Grid Adjustment Factor before using them to inverse between the measured points for the final survey.

GRID ADJUSTMENT FACTOR

Each survey will require a project Grid Adjustment Factor (GAF) (a.k.a. combined factor) specific for the location within the State Plane Zone (SPZ). The distance from the Central Meridian (Transverse Mercator Projection) or Central Parallel (Lambert Conformal Conic Projection) will determine its value.

The GAF is a combination of the Grid Scale Factor and the Ellipsoid (Elevation) Factor. Typically only one GAF will be established for the survey project. Large projects may be divided into smaller areas each with their own GAF depending on the amount of distortion across the overall project area.

The GAF is used to convert Grid Coordinates to Ground Coordinates and vice versa.

It is important to know where your project is located within the SPZ. For most of the SPZ, the Grid plane will be below the surface therefore Grid distances will be shorter than Ground distances. When out on the fringes of the SPZ, the Grid plane can be above the surface as shown below.

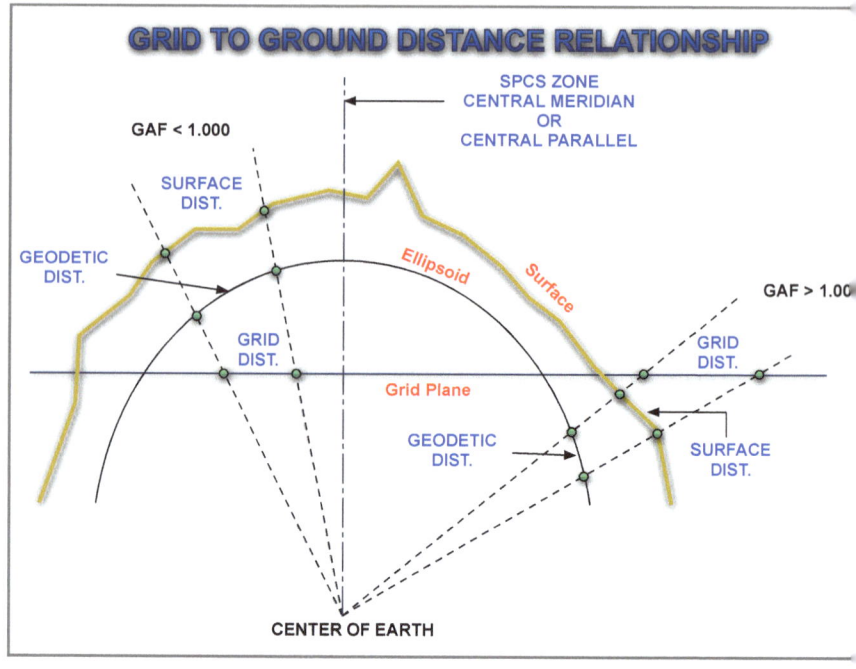

The GAF can be shown in two-forms on surveys and data sheets.

The GAF can be shown in the format as shown on the NGS Data Sheet such as 0.99985086 which correlates to the image above. The reciprocal value (1/0.99985086) of 1.00014916 is also used. The format that is used will depend on the client requirements and/or the surveyors preference. Be sure to verify the GAF before applying it to any coordinat conversion. It is very easy to apply the GAF format incorrectly to convert between Grid to Ground coordinates.

There are two methods to apply the GAF to Grid Coordinates; scale about the origin of (0,0) or a point locate within the project area.

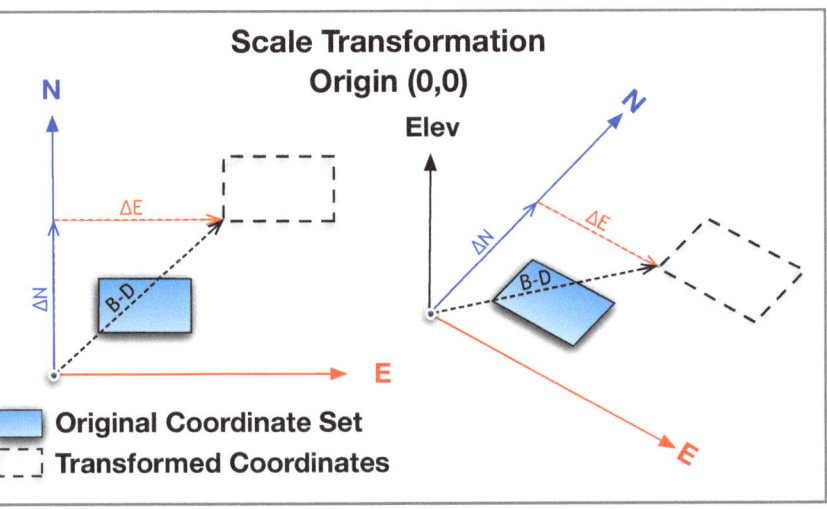

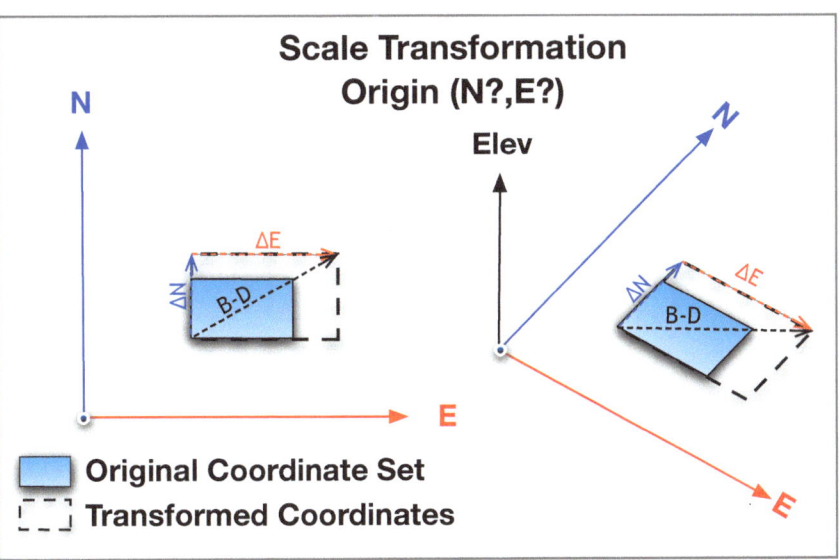

The method most widely used is to scale about origin (0,0) as follows:

Grid_northing x GAF = Ground_northing

Grid_easting x GAF = Ground_easting

Note: The GAF is in the reciprocal form (1/x).

It is recommended you use five decimal places for the coordinates so that when you inverse between a pair of coordinates, the bearing and distance are more accurate. Rounding to the nearest second for the bearing and two decimal places for the distance will have a significant impact for short distances when using only two decimal places for the coordinates.

It is recommended that when applying Grid to Ground functions in software applications, that a sanity check be made using manual calculations on a couple of random points.

A Grid Adjustment Factor app is available for the iOS and Android smart devices that can be used to perform these sanity checks.

http://www.cc4w.net/apps.html

See the scale transformation example at the end of the book.

See "Coordinate Transformation - Book 9" for formulas and more details on these scaled transformations.

NGS DATA SHEET

Below is a typical NGS Data Sheet for a Vertical Control
oint. This data sheet has a lot of information on it. The area
the red boxes (*Orthometric Height, Geoid, Vertical Order*)
is the only information that is required on this sheet. The
Latitude and Longitude are approximate and should not be
used.

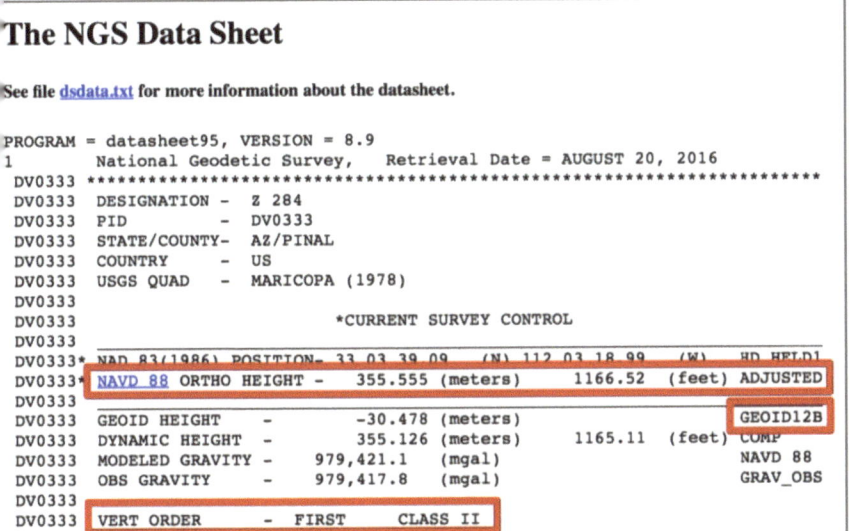

The Geoid is used to convert the ellipsoidal height obtained
from the Global Navigation Satellite System to orthometric
height of specific vertical datum.

Below is a typical NGS Data Sheet for a Horizontal Control
oint. This data sheet has a lot of information on it. The area
in the red boxes (*Latitude, Longitude, Grid Northing, Grid
Easting, Convergence angle, Combined Factor*) is the only
information that is required. The vertical information is
approximate and should not be used.

The NGS Data Sheet

See file dsdata.txt for more information about the datasheet.

```
PROGRAM = datasheet95, VERSION = 8.9
1          National Geodetic Survey,   Retrieval Date = AUGUST 20, 2016
 AJ3728 ********************************************************************
 AJ3728  HT_MOD      -  This is a Height Modernization Survey Station.
 AJ3728  DESIGNATION -  1FD1
 AJ3728  PID         -  AJ3728
 AJ3728  STATE/COUNTY-  AZ/MARICOPA
 AJ3728  COUNTRY     -  US
 AJ3728  USGS QUAD   -  GLENDALE (1982)
 AJ3728
 AJ3728                       *CURRENT SURVEY CONTROL
 AJ3728
 AJ3728* NAD 83(2011) POSITION- 33 35 50.86841(N) 112 08 57.87395(W)   ADJUSTED
 AJ3728* NAD 83(2011) ELLIP HT-   349.664 (meters)       (06/27/12)    ADJUSTED
 AJ3728* NAD 83(2011) EPOCH   -  2010.00
 AJ3728* NAVD 88 ORTHO HEIGHT -   379.46  (meters)    1244.9   (feet) GPS OBS
 AJ3728
```

```
 AJ3728. The following values were computed from the NAD 83(2011) position.
 AJ3728
 AJ3728;                    North         East      Units Scale Factor Converg.
 AJ3728;SPC AZ C      -  288,033.737   191,759.355   MT  0.99990575   -0 07 43.6
 AJ372 ;SPC AZ C      -  944,992.58    629,131.74    iFT 0.99990575   -0 07 43.6
 AJ3728;UTM   12      -  3,718,118.962 393,333.678   MT  0.99974021   -0 38 09.9
 AJ3728
 AJ3728!              -  Elev Factor  x  Scale Factor =  Combined Factor
 AJ3728!SPC AZ C      -  0.99994511   x  0.99990575   =  0.99985086
 AJ3728!UTM   12      -  0.99994511   x  0.99974021   =  0.99968533
 AJ3728
 AJ3728                       SUPERSEDED SURVEY CONTROL
 AJ3728
 AJ3728  NAD 83(2007)-  33 35 50.86792(N)    112 08 57.87427(W) AD(2007.00) 0
 AJ3728  ELLIP H (02/10/07)  349.675  (m)                       GP(2007.00)
 AJ3728  NAD 83(1992)-  33 35 50.86716(N)    112 08 57.87352(W) AD(       ) B
 AJ3728  ELLIP H (04/12/01)  349.710  (m)                       GP(       ) 3 2
 AJ3728  NAVD 88 (04/12/01)  379.47   (m)  GEOID99 model used   GPS OBS
 AJ3728
```

The information noted as (2011) is the datum tag that refers to the year the control point position was completed and the North American tectonic plate to which the coordinates are referenced.

Note: This date is often referred to as the 2011 Epoch which is technically incorrect. Per this data sheet the Epoch is actually 2010.00.

When two or more NGS Control Points are used on the same project, the datum tag (*year*) must be the same year. You cannot mix the Meta Data between different datum tags (*year*).

The Superseded Survey Control gives the values of earlier datum tag (*year*) which may be needed when retracing older surveys that used those datums.

Whenever possible, the latest datum tag (*year*) should be used.

The combined factor (GAF) shown on the NGS Data Sheet is for Ground to Grid coordinate conversion when using the following formulas:

Ground_northing x GAF = Grid_northing

Ground_easting x GAF = Grid_easting

or Grid to Ground

Grid_northing / GAF = Ground_northing

Grid_easting / GAF = Ground_easting

Note: Scale origin (0,0)

OPUS SOLUTION

Below is an OPUS Project Solution for a Horizontal and Vertical Control Point. This data sheet has a lot of information on it. The areas in the red boxes is the only information that is required on this sheet.

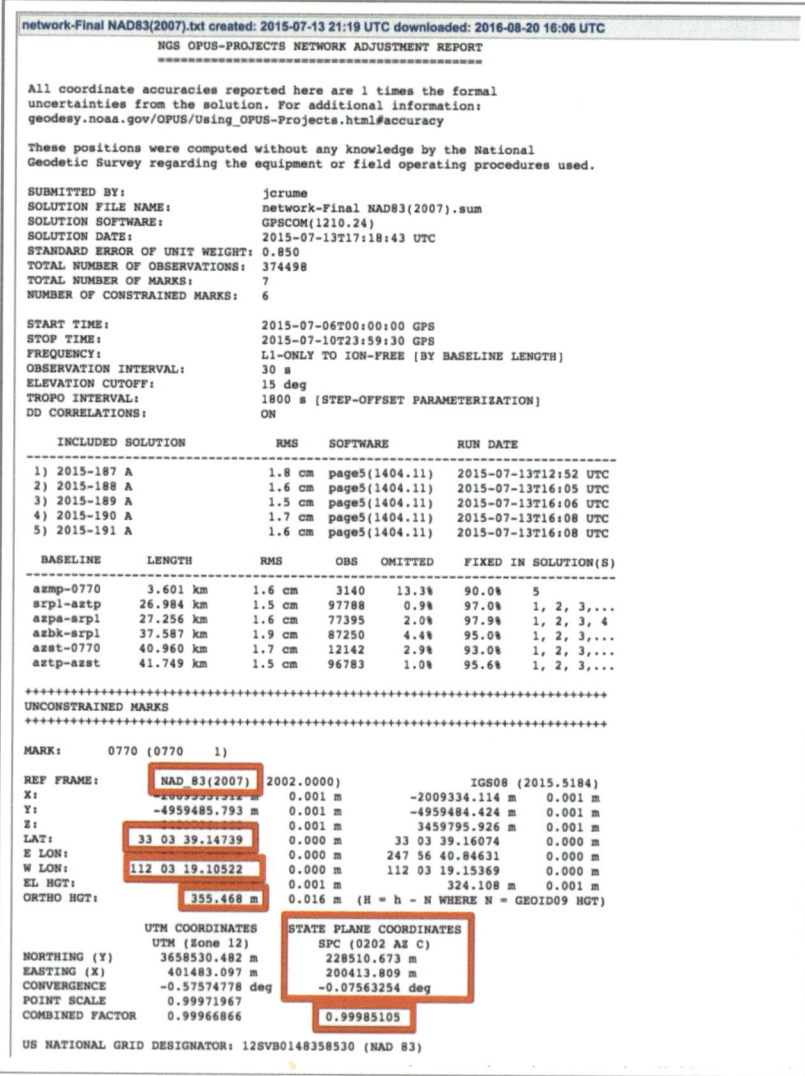

```
MARK:        0770 (0770     1)

REF FRAME:       NAD_83(2007) (2002.0000)                          I
X:               -2009333.312 m       0.001 m           -2009334.
Y:               -4959485.793 m       0.001 m           -4959484.
Z:                3459796.065 m       0.001 m            3459795.
LAT:            33 03 39.14739         0.000 m        33 03 39.16
E LON:          247 56 40.84574        0.000 m        247 56 40.84
W LON:          112 03 19.10522        0.000 m        112 03 19.15
EL HGT:             324.294 m          0.001 m               324.
ORTHO HGT:          355.468 m          0.016 m   (H = h - N WHERE

                UTM COORDINATES       STATE PLANE COORDINATES
                UTM (Zone 12)            SPC (0202 AZ C)
NORTHING (Y)     3658530.482 m          228510.673 m
EASTING (X)       401483.097 m          200413.809 m
CONVERGENCE      -0.57574778 deg       -0.07563254 deg
POINT SCALE       0.99971967            0.99990207
COMBINED FACTOR   0.99966866            0.99985105
```

The combined factor (GAF) shown on the OPUS Solution is for Ground to Grid coordinate conversion when using the following formula:

Ground_northing x GAF = Grid_northing

Ground_easting x GAF = Grid_easting

or Grid to Ground

Grid_northing / GAF = Ground_northing

Grid_easting / GAF = Ground_easting

Note: Scale origin (0,0)

NEW DATUMS ARE COMING

NOAA is Replacing NAD 83 and NAVD 88. NOAA's Nationa Geodetic Survey (NGS) will be replacing the datums of the National Spatial Reference System (NSRS) in year 2022, including the North American Datum of 1983 (NAD 83) and the North American Vertical Datum of 1988 (NAVD 88). NGS will provide the tools to easily transform between the new and old datums.

SCALE TRANSFORMATION EXAMPLE

$$Grid\_northing \times GAF = Ground\_northing$$

$$Grid\_easting \times GAF = Ground\_easting$$

GRID TO GROUND EXAMPLE				
	NGS	1/X	Scale Origin	
Project GAF	0.99985086	1.00014916	N = 0, E = 0	
	Grid Northing	Grid Easting	Ground Northing	Ground Easting
AJ3728	944992.58000	629131.74000	945133.53722	629225.58270
100	944592.32615	629189.12356	944733.22366	629282.97482
101	944192.07230	629246.50712	944332.91011	629340.36694
102	943791.81845	629303.89068	943932.59656	629397.75906
103	943391.56460	629361.27424	943532.28300	629455.15118

The above formulas and table uses the reciprocal of 0.99985086 (NGS Combined Factor) for the GAF. This is th format that I have used for many years and it is my format preference.

The scale origin is (0,0) which makes the scaling transformation easy and clean. This is the preferred method

Be aware that scaling about a specific point other than (0,0) is used by several agencies and surveyors that prefer the more complex method.

 See "Coordinate Transformation - Book 9" for formulas and more details on these scaled transformations.

 Download the free Windows CleanAscii-Pro software to perform the Grid to Ground and vice versa on an ASCII file at "www.cc4w.net/freestuff.html".

USEFUL LINKS

NGS Data Sheets: http://www.ngs.noaa.gov/datasheets/
OPUS: http://www.ngs.noaa.gov/OPUS/
Geodetic Datums: https://youtu.be/kXTHaMY3cVk

Author's Website: http://www.cc4w.net/index.html
Reference Books: http://www.cc4w.net/ebooks.html
Surveying Apps: http://www.cc4w.net/apps.html
Freestuff: http://www.cc4w.net/freestuff.html
Study Guide: http://www.cc4w.net/study-guide.html

Note: *Links do become broken over time. If the above links do not work, you will need to do an online search to find the new links.*

CONCLUSION

The datum that you use on your survey will be dependent upon your client requirements and/or personal preferences.

Coordinates on a drawing, plat or ASCII file need to be identified as ground or grid. A project GAF needs to be shown as well.

The Basis of Bearing needs to be identified as Grid if a GPS was used. Distances should be Ground and identified as such unless Grid is required by the client.

The only way to protect your survey is to document it thoroughly so that other surveyors can follow in your footsteps. Lack of documentation can become a real issue especially if calibration/localization is used instead of a simple scaling transformation using a GAF from a single control point. Calibration/localization will be covered in detail in future books in this series.

Remember that all of the points in a survey are relative to each other regardless of the datum it is on. Changing from one datum to another will have no affect on the relativity of the points with each other.

Some technicians and professionals tend to *freak-out* when it comes to the horizontal and vertical datum and forget all about the relativity. They also make using State Plane Coordinates more difficult than it really is. When you use the correct map projection and GAF, it becomes a simple scaling transformation to go from Grid to Ground and vice versa.

Communication between the field and office personnel is paramount. A seamless operation is required to bring a project to successful completion.

ABOUT THE AUTHOR

Jim Crume P.L.S., M.S., CFedS

My land surveying career began several decades ago while attending Albuquerque Technical Vocational Institute in New Mexico and has traversed many states such as Alaska, Arizona, Utah and Wyoming. I am a Professional Land Surveyor in Arizona, Utah and Wyoming. I am an appointed United States Mineral Surveyor and a Bureau of Land Management (BLM) Certified Federal Surveyor. I have many years of computer programming experience related to surveying.

This ebook is dedicated to the many individuals that have helped shape my career. Especially my wife Cindy. She has been my biggest supporter. She has been my instrument person, accountant, advisor and my best friend. Without her, would not be the professional I am today. Cindy, thank you very much.

Other titles by this author:

http://www.cc4w.net/ebooks.html

Follow us on Facebook

Books available on amazon.com